D0096357

THE AUDUBON SOCIETY POCKET GUIDES

A Chanticleer Press Edition

Ann H. Whitman
Editor

Jerry F. Franklin
John Farrand, Jr.
Consultants

Eastern Region

FAMILIAR TREES OF NORTH AMERICA

Alfred A. Knopf, New York

This is a Borzoi Book
Published by Alfred A. Knopf, Inc.

Prepared and produced by Chanticleer Press, Inc.,
New York.
Color reproductions by Nievergelt Repro AG,
Zurich, Switzerland.
Typeset by Dix Type Inc., Syracuse, New York.
Printed and bound by Dai Nippon, Tokyo, Japan.

Published October 1986
Reprinted once
Third Printing, September 1989

Library of Congress Catalog Number: 86-045585
ISBN 0-394-74851-4

Contents

How to Use This Guide

Trees are our largest and most conspicuous plants, a dominant part of nearly every landscape. Learning to identify trees will not only add to your appreciation of the American countryside, but will also serve as a good introduction to other aspects of nature, since trees usually have a great influence on the kinds of birds, wildflowers, mammals, and other wildlife in an area.

Coverage
This new guide covers 80 of the most common and frequently encountered tree species in the East. Our range is bounded by the Arctic tree line across northern Canada, the Atlantic Ocean on the east, the Gulf of Mexico to the south, and the Rocky Mountains on the west. (Thus, the boundary roughly follows the Rockies.) The companion volume to western trees covers species west of this boundary.

Organization
This easy-to-use guide is divided into three parts: introductory essays, illustrated accounts of the trees, and appendices.

Introduction
As a basic introduction, the essay "Identifying Trees" suggests questions to ask yourself when you see an unfamiliar tree. "Key Features of a Tree" describes and illustrates a tree's characteristic elements—leaf shapes,

fruits and cones, and representative silhouettes. An awareness of these features is essential to identification.

The Trees   This section contains 80 color plates arranged visually by the overall shape of a tree and its leaf. The first group includes broadleaf hardwoods and the second, coniferous trees. Facing each color plate is a description of the tree's most important features, such as its flowers and fruit, habitat, geographic range, and elevation. The introductory paragraph for each species discusses the uses of the wood, folklore, and other subjects. A black-and-white silhouette of the tree supplements the photograph. For evergreens, the silhouette shows the tree's year-round appearance; for trees that shed their leaves, winter silhouettes show the basic tree shape. A close-up of the bark—often an important identifying feature—is also included with each account.

Appendices   Featured here is an essay on the 20 common families of trees in our area. Knowing family traits helps to recognize many related species.

Whether you live in the country, surrounded by peaceful woods, or in the city, this guide is certain to bring you pleasure and a deeper understanding of nature.

7

Identifying Trees

The most important step in identifying trees is learning what features to look for. This "questionnaire" is designed to help teach you the points to consider when you look at a tree; in a step-by-step fashion, you will go from the general to the specific and soon narrow the identification to a few clear choices. Remember to take into account the habitat and range of a tree as well as its physical characteristics when you read the text descriptions in the book.

Conifers

If the tree is a conifer, are the leaves shaped like needles (pines, firs, spruces, larches, and others), scales (cypresses), or awls (junipers)? If needlelike, are they borne in clusters (pines, larches) or singly? If in clusters, how many are in a bunch? If the needles are borne singly (hemlocks, spruces, firs, Douglas-firs), are they sharply pointed or rounded on the end?

Once you have examined the needles, look at the cones. Are they upright or pendent? Small or large? Where on the tree are they growing?

Broadleaf Hardwoods

If the tree is a hardwood, are the leaves simple or compound? If they are simple, are they opposite (maples), or alternate (oaks, elms, poplars)? Are the leaves lobed? Are the lobes palmate (maples,

8

Sweetgum?) Or pinnate (oaks)? If the leaves are not lobed, are they toothed (elms, poplars, birches)? Are the leaves evergreen (live oaks, rhododendrons, magnolias)? Do they have spiny margins (hollies)? Is the bark deeply furrowed (Yellow-poplar, many oaks), smooth and green or white (poplars), or papery and peeling (birches)? Are there large, showy flowers (magnolias, rhododendrons, Yellow-poplar)? What kind of fruit is present? A winged key (maples)? A dense ball of seeds (Sweetgum, sycamores)? A conelike cluster of seeds (birches)? An acorn (oaks)? A stone fruit (cherries, hawthorns)?

If the leaves are compound, are they pinnately compound (ashes, sumacs, locusts, hickories) or palmately compound (buckeyes)? If the leaves are pinnately compound, are they opposite (ashes) or alternate (hickories, locusts)? Do the stems have thorns (locusts)? What kind of fruit is present? Is the plant shrubby (sumacs)? Is the tree growing in a swamp and does it have a swollen or fluted base (baldcypresses, tupelos)?

Key Features of a Tree

In learning to identify trees, it is important to become acquainted with the major elements of each species—its leaf shape, fruit or cones, and silhouette. The pages that follow present examples of the diversity of these significant characteristics.

Leaf Shapes

Recognizing leaf shapes at a glance is a quick way to become comfortable with identifying trees. In some cases, family members have similarly shaped leaves; for example, if you can recognize the classic "maple leaf," you will be able to identify not only all the maples in this book but also a variety of similar relatives. On the pages that follow, 10 of the most common leaf shapes are illustrated for you.

Fruits and Cones

Another good way to recognize a tree is to learn what its fruit looks like. This is especially useful for some deciduous trees in late fall or early winter, when the fruit may persist after the leaves have fallen. If leaves are still present, you can use the fruit to confirm your identification. In some large groups, such as the oaks, identifying the fruit may be the simplest and surest way to distinguish similar species.

The fruit of a tree is a developed, fertilized ovary that contains the seed or seeds. Because fruits develop from flowers, they reflect, in their arrangement, the arrangement of the flowers of that species. Thus trees that bear solitary flowers, such as magnolias, also bear solitary fruits; those species that bear clusters of flowers, such as hollies, also have clusters of fruits.

Some fruits, like cherries or elm keys, are one-seeded; others, such as mulberries or the fruits of magnolias, contain hundreds of seeds. Conifers bear their seeds within cones, which differ from the fruits of broadleaf trees in several important scientific ways. Conifer cones typically contain hundreds of seeds.

A tree may bear nuts, berries, drupes, capsules, or some other kind of fruit. On pages 14 and 15, you will find illustrated 10 different fruit types.

Silhouettes   Like learning the shape of a leaf or the kind of fruit a tree bears, recognizing its overall shape, or silhouette, is often a means to identification, for many trees have a characteristic shape. On pages 16 and 17, you will find illustrations of five typical silhouettes—winter shapes for deciduous trees and year-round silhouettes for evergreens—that will introduce you to the principles of recognizing a tree by its shape.

Leaf Types

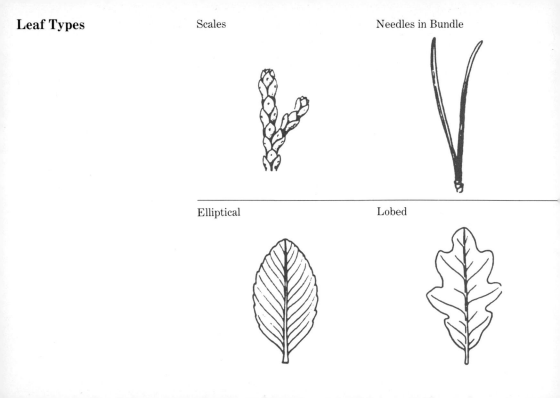

Scales

Needles in Bundle

Elliptical

Lobed

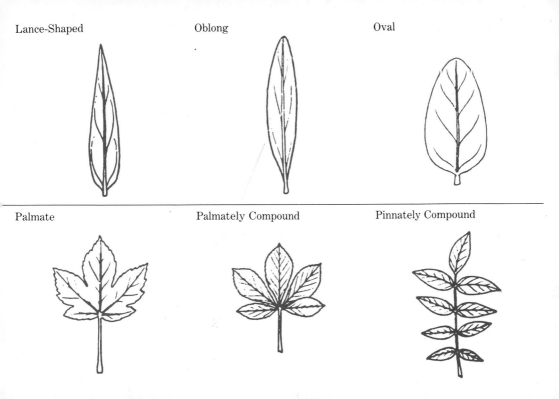

Lance-Shaped

Oblong

Oval

Palmate

Palmately Compound

Pinnately Compound

Types of Fruits and Cones

Cone
pines, firs, hemlocks, spruces, larches

Nut
chestnuts, hickories, beeches, buckeyes

Pod
locusts, acacias

Double key (samara)
maples

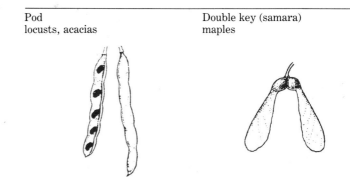

Acorn
oaks, Tanoak

Capsule
poplars, willows, bays, Mountain-laurel

Multiple of nutlets
Sycamore, Sweetgum, birches, Yellow-poplar

Drupe
cherries, plums, hawthorns, peaches

Berry
hollies, dogwoods, elders

Pome
apples, pears

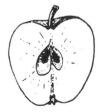

Tree Silhouettes

Narrow, pointed crown of spreading branches

Straight, narrow crown of horizontal branches

Open, rounded crown

Rounded crown of widely
spreading branches

Narrow, open, rounded crown

THE TREES

Pussy Willow *Salix discolor*

In late winter and early spring, the flower buds of the Pussy Willow open, revealing the soft, silky hair that gives this species its common name. In winter, Pussy Willow flowers can be forced indoors by placing cut twigs in water; some flowers have golden stamens, and others greenish pistils.

Identification   Height: 20′; diameter: 8″. Shrub or small tree with many stems and an open, rounded crown. Leaves shiny green above, whitish below; 1½–4½″ long, ⅜–1¼″ wide; lance-shaped or narrowly elliptical, with wavy-toothed edges. Flowers tiny, in compact clusters or catkins 1–2¼″ long; catkins cylindrical, with blackish scales, covered with silky "fur" in late winter. Fruit a narrow capsule, ⁵⁄₁₆–½″ long; light brown with fine hairs; appearing before leaves in spring.

Habitat   Streamsides, lakeshores, and wet meadows; often in conifer forests.

Range   NE. Canada and New England south to Delaware, west to British Columbia and North Dakota; to 4000′.

Black Willow *Salix nigra*

The largest and most important willow in North America, this species is found throughout the East. The wood is used for furniture, toys, barrels, and pulpwood. Large Black Willows help prevent erosion along stream banks. A subspecies in the lower Mississippi Valley reaches 120 feet or more.

Identification Height: 60–100′; diameter: 1½–2½′. Large tree with 1 or more trunks, usually straight and leaning; upright branches form narrow or irregular crown. Leaves shiny green above, paler below; 3–5″ long and narrowly lance-shaped, often curved to one side; pointed, with finely saw-toothed edges. Flowers tiny, in compact clusters 1–3″ long, with yellow, hairy scales; appearing at ends of twigs in spring. Fruit a reddish-brown capsule, ³⁄₁₆″ long, maturing in late spring.

Habitat Wetland areas, along stream banks and lakeshores, and especially in floodplains; in pure stands or with cottonwoods.

Range Great Lakes to Atlantic Coast, south through Mississippi Valley to Gulf Coast; to 5000′.

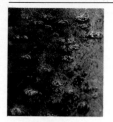

Southern Bayberry *Myrica cerifera*

The Southern Bayberry—also called Southern Waxmyrtle—is a popular ornamental throughout the Southeast. Its berrylike fruit has a waxy covering that can be removed in boiling water; in Colonial times, fragrant candles were made from this material, and bayberry candles are still popular today.

Identification	Height: 30'; diameter: 6". Evergreen shrub or small tree with narrow, rounded crown. Leaves shiny yellowish green above, with tiny dark brown dots; paler below, often hairy, with orange dots; 1½–3½" long and ¼–¾" wide; aromatic. Flowers tiny, yellow-green, in cylindrical clusters about ¼–¾" long; appearing at base of leaf in early spring. Fruit berrylike, light green, and warty; covered with bluish-white wax and appearing in clusters; maturing in fall and remaining through the winter.
Habitat	Moist, sandy soil; along stream banks and in swamps, pinelands, and upland hardwood forests.
Range	S. New Jersey to S. Florida, west to S. Texas, and north to SE. Oklahoma; to about 500'.

24

Rosebay Rhododendron *Rhododendron maximum*

This large, hardy evergreen rhododendron forms dense thickets. It is found throughout the Appalachians and is especially abundant in the Great Smoky Mountains National Park. Because of its large white blooms, it is also a popular ornamental. Honey made from rhododendrons is poisonous.

Identification Height: 20'; diameter: 6". Evergreen shrub or small tree with short, crooked trunk and broad, rounded crown; the branches are stout and crooked. Leaves shiny dark green above, whitish below with fine hairs; 4–10" long and 1–3" wide, pointed at each end, and often with curled edges. Flowers white, clustered; each blossom 1½" wide, with 5 rounded, petal-like lobes forming a bell-shaped corolla, the largest or upper lobe with many green spots; appearing in summer. Fruit a small, oval, reddish-brown capsule, about ½" long, on a long stalk and covered with fine hairs; maturing in fall, remaining through winter.

Habitat Moist soils, especially along streams in mountain forests.

Range Throughout the Northeast, mostly in mountains; in Appalachians to Georgia; to 6000' in the South.

26

Mountain-laurel *Kalmia latifolia*

In spring, this beautiful shrub bursts forth in a glory of blossoms. Mountain-laurel is a member of the heath family, to which azaleas and rhododendrons also belong. This species forms dense thickets, called heath balds or laurel slicks, in the southern Appalachians. Honey made from the flowers is probably poisonous.

Identification Height: 20′; diameter: 6″. Evergreen shrub with many stems or small tree with a short, crooked trunk; the branches are stout and spreading, forming a compact, rounded crown. Leaves thick and stiff; dark, dull green above, yellow-green below; about 2½–4″ long, 1–1½″ wide, narrowly elliptical, with smooth edges and a hard, whitish point at the tip. Flowers white or pink, in clusters; each bloom ¾–1″ wide, with 5 petal-like lobes forming a saucer-shaped corolla; appearing in spring. Fruit a round, hairy, dark brown capsule, ¼″ wide, with a long "thread" at the tip, maturing in fall and persisting.

Habitat Acid soils in mixed upland forests or valleys.

Range S. Maine to N. Florida, west to Indiana and Louisiana; to 4000′ (higher in southern mountains).

Sweetbay *Magnolia virginiana*

The fragrant flowers of this attractive magnolia are produced over a long period in the late spring and early summer, making the Sweetbay popular as an ornamental. It is also known as Swampbay and Swamp Magnolia. In the North, it sheds its leaves in winter, but it is almost evergreen in the South.

Identification Height: 20–60'; diameter: 1½'. Tree with narrow, rounded crown and aromatic, spicy leaves and twigs. Leaves shiny green above, whitish below with fine hairs; 3–6" long, 1¼–2½" wide; thick, oblong, with a blunt tip and smooth edges. Flowers large, white, cup-shaped; 2–2½" wide, with 9–12 white petals; fragrant. Fruit a conelike, dark red cluster of many separate fruits, each pointed; 1½–2" long; maturing in early fall.

Habitat Wetland areas: coastal swamps, pond shores, and streamsides.

Range Long Island to S. Florida, west to SE. Texas; local in NE. Massachusetts; to 500'.

Cockspur Hawthorn *Crataegus crus-galli*

This hawthorn is easy to identify because of its very long spines, which grow on both bark and twigs, and its shiny, dark green, spoon-shaped leaves. Common and widely distributed, the Cockspur Hawthorn has long been a popular ornamental; in the fall, its scarlet leaves and berries are bright and colorful.

Identification   Height: 30'; diameter: 1'. Small, thicket-forming tree with short, stout trunk; the branches are spreading and horizontal, forming a dense crown. Leaves shiny dark green above, paler and veined below; 1–4" long, ⅜–2" wide; spoon-shaped, with a rounded tip or short-pointed; sharp, gland-tipped sawteeth beyond middle; slightly thick and leathery. Flowers in large white clusters; each bloom ½–⅝" wide, with 5 white petals; appearing in spring or early summer. Fruit berrylike, in clusters; greenish or dull, dark red; ⅜–½" in diameter; maturing in late fall and persisting until spring.

Habitat   Moist soils in valleys or low upland slopes.

Range   SE. Canada to N. Florida, west to E. Texas, and north to Iowa; to 2000'.

Live Oak *Quercus virginiana*

The image of this evergreen oak, with its branches draped in pale gray-green Spanish Moss, is a familiar emblem of the South. The tree is a popular ornamental in the Southeast, where it grows to large size. Live Oak timber was once important in this nation's shipbuilding industry.

Identification Height: 40–50′; diameter: 2–4′. Medium-size evergreen tree with a short, broad trunk, buttressed at the base and forking into nearly horizontal branches, and with a very broad, spreading crown. Leaves shiny dark green above, gray-green below with dense hairs; 1½–4″ long, ⅜–2″ wide; oblong with a rounded tip, sometimes with a tiny tooth; edges usually smooth, sometimes curled under slightly. Acorns oblong, ⅝–1″ wide; one-fourth to half enclosed by a deep cup; green, becoming brown, and long-stalked; maturing during the first year.

Habitat Sandy soils and coastal areas near marshes; often in pure stands.

Range SE. Virginia to S. Florida, west to S. Texas; local in SW. Oklahoma and NE. New Mexico; to 300′ in most areas.

34

Southern Magnolia *Magnolia grandiflora*

This ornamental and shade tree is popular everywhere in warm temperate and subtropical regions. It is one of our most beautiful native trees, producing lovely, fragrant, large white flowers in late spring and early summer; both Louisiana and Mississippi have adopted these luscious blooms as their state flower.

Identification   Height: 60–80′; diameter: 2–3′. Evergreen tree, with a straight trunk and pointed crown. Leaves shiny bright green above, pale with rust-colored hairs below; 5–8″ long and 2–3″ wide; oblong, thick, with edges slightly turned under. Flowers large, cup-shaped; 6–8″ wide, with 6 or more petals; appearing singly at end of twig. Fruit conelike, 3–4″ long; oblong and pink or brown, made up of many smaller fruits.

Habitat   Moist valleys and low uplands, often with other hardwoods.

Range   E. North Carolina south to central Florida, west to E. Texas; hardy as far north as Philadelphia in parks and gardens; to 400′.

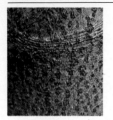

American Holly *Ilex opaca*

The American Holly is widely grown as an ornamental, and its foliage is familiar as a Christmas decoration. The wood is whitish, with a fine texture, and is used for inlays in cabinetry and carvings. The related English Holly (*I. aquifolium)* is also widespread in the East; it has shinier leaves and more plentiful berries.

Identification
Height: 40–70′; diameter: 1–2′. Evergreen tree with a dense, rounded, narrow crown. Leaves dark green above, yellow-green below, with spines; 2–4″ long and ¾–1½″ wide; elliptical with spiny points and coarse, spiny teeth. Flowers white, with 4 petals; ¼″ wide; in short clusters, appearing at base of new leaves in spring; male and female on different trees. Fruit bright red, berrylike; ¼–⅜″ diameter; maturing in fall, remaining through the winter.

Habitat
Floodplains and other moist or wet, well-drained areas; in mixed hardwood forests.

Range
E. Massachusetts to central Florida, south-central Texas, and Missouri; to 4000′ in most areas, higher in southern Appalachians.

American Chestnut *Castanea dentata*

Formerly abundant in the East, this once-stately tree reached a height of 100 feet. But it fell victim to an introduced fungus in 1904, and within 40 years was nearly wiped out. Today, sprouts continue to grow from surviving roots of former trees, but within a short time the sprouts are killed back by the blight. In some parts of the country, however, the fungus is absent, and cultivated Chestnuts continue to thrive.

Identification Height: 20′; diameter: 4″. Small sprouts growing to a moderate height. Leaves shiny yellow-green above, paler below; turning yellow in fall; narrowly oblong and pointed; 5–9″ long and 1½–3″ wide, with many straight parallel side veins and a curved tooth at the tip. Flowers in clusters, 6–8″ long at base of leaf; many tiny whitish male flowers, fewer greenish female flowers. Fruit a bur, 2–2½″ wide, with many spines; maturing in fall; edible chestnuts within.

Habitat Moist upland soils in mixed forests.

Range SE. Canada and Maine to Georgia, Indiana, and Mississippi; to 4000′.

40

Slippery Elm *Ulmus rubra*

The gluey, slightly aromatic inner bark gives this elm its common name. This "slippery" layer of bark is edible, and when first dried and then moistened can be used as a cough medicine or poultice. Spring rains may cause the pigment of the flower buds to run, giving the whole tree a reddish hue.

Identification Height: 70′; diameter: 2–3′. Large tree with broad crown of spreading branches. Leaves green or dark green above and covered with soft hairs below; 4–7″ long, in 2 rows; elliptical, rough, and doubly saw-toothed, with many straight parallel veins; turning dull yellow in fall. Flowers greenish, ⅛″ long, numerous; appear on short stalks along twigs in early spring. Fruit a nearly round, flat key (samara), ½–¾″ long, with a light green, broad, hairless wing, slightly notched at the tip; maturing in spring.

Habitat Floodplains and lower slopes with moist soils; also often on dry uplands in hardwood forests.

Range S. Ontario to S. Quebec and Maine, south to NW. Florida and central Texas, west to SE. North Dakota; to 2000′.

American Elm *Ulmus americana*

This large, handsome, graceful tree is a familiar sight along city streets and in gardens. Once abundant throughout the East, its numbers have dwindled significantly since 1930, when the species fell prey to an introduced fungus. The wood of this and other elms is used for a variety of purposes.

Identification Height: 100'; diameter: 4'. Trunk often with buttresses at base; many spreading branches form a broad, rounded crown, often wider than high. Leaves dark green above, usually hairless; paler below, with soft hairs; 3–6" long and 1–3" wide, in 2 rows; elliptical, abruptly long-pointed, rounded at the base with unequal sides; edges doubly saw-toothed; turning bright yellow in fall. Flowers greenish, ⅛" long, in clusters along twigs; in early spring. Fruit an elliptical, flat, winged key, ⅜–½" long; with points curving inward; maturing in early spring.

Habitat Valleys, floodplains, and mixed hardwood forests.

Range Saskatchewan east to Atlantic Coast, south to central Florida and Texas; to 2500'.

44

Eastern Hophornbeam *Ostrya virginiana*

The very tough wood of this species is used for making tool handles and fences, and inspired the alternate name, Ironwood. The Eastern Hophornbeam, which is often used as an ornamental, gets its name from the conelike fruit clusters, which resemble hops. In the wild, these nuts are consumed by various animals.

Identification Height: 20–50′; diameter: 1′. Tree with rounded crown of slender, spreading branches above a muscular-looking trunk. Leaves dull yellow-green, nearly hairless above; paler below with hair on the veins; 2–5″ long and 1–2″ wide; elliptical with double saw-toothed edges and many straight parallel side veins; turning yellow in fall. Flowers tiny; male flowers greenish, 1½–2½″ long, in cylindrical, drooping clusters; female flowers reddish green, in narrow cylindrical clusters ½–¾″ long; appearing before leaves. Fruit a conelike cluster, 1½–2″ long, made up of many small brown nutlets with papery coverings.

Habitat In understory of upland hardwood forests; in moist soils.

Range SE. Manitoba to Atlantic Coast, south to N. Florida, and west to E. Texas; to 4500′.

46

American Beech *Fagus grandifolia*

This handsome shade tree is widespread in the East, where its nuts provide food for a variety of wildlife. Beeches are unusual because their bark remains smooth, even on old trees. The wood of this species is used chiefly for furniture, floors, and interior trim. Colonies of these trees grow from root suckers.

Identification
Height: 60–80'; diameter: 1–2½'. Large tree with long, spreading branches and a rounded crown. Leaves dull, dark blue-green above, lighter below, with few or no hairs; 2½–5" long and 1–3" wide; elliptical, long-pointed at tip, with coarse, saw-toothed edges; turning yellow and brown in fall. Male flowers small, yellowish; in a delicate, tassel-like cluster, ¾–1" in diameter. Female flowers in pairs at end of short stalk; ¼" long, with narrow, hairy, reddish scales. Fruit a prickly brown bur, ½–¾" long; maturing in fall and dividing into 4 parts; usually containing 2 shiny nuts.

Habitat
Uplands and well-drained lowlands with rich, moist soil.

Range
S. Ontario to Atlantic Coast, south to N. Florida and E. Texas; to 3000' (6000' in southern Appalachians).

48

Sweet Birch *Betula lenta*

Widely known as Black Birch, this species produces sap that can be tapped, fermented, and made into birch beer. The crushed twigs and foliage are aromatic. Birch oil (oil of wintergreen) can be obtained from young trees, although today most wintergreen flavoring is produced artificially.

Identification Height: 50–80'; diameter: 1–2½'. Tree with spreading branches forming a rounded crown. Leaves: 2½–5" long, 1½–3" wide; dull dark green above, lighter yellow-green below; elliptical, long-pointed, and often with a notch at the base; edges have sharp double sawteeth; usually with 9 to 11 pairs of parallel side veins; turning bright yellow before shedding in fall. Flowers tiny; male flowers yellowish, in drooping catkins 3–4" long; female flowers greenish, in short, erect catkins; appearing in early spring. Cones brownish, oblong; ¾–1½" long; containing winged nutlets; maturing in fall.

Habitat In moist, cool upland areas; with conifers and hardwoods.

Range S. Maine to Ohio and N. Alabama; 2000–6000' in southern mountains, lower altitudes northward.

European Buckthorn *Rhamnus cathartica*

Introduced from Europe, where it has long been a favorite ornamental, the European Buckthorn is often used in this country as a hedge. The scientific name refers to the fact that the berries were once used as a cathartic. The brown outer bark of this species peels gradually to reveal a reddish layer of inner bark.

Identification   Height: 20′; diameter: 4″. Spiny shrub or small tree. Leaves green above, paler below; ¾–2½″ long, ¾–2″ wide; broadly elliptical, with a distinctive pattern of curved veins; edges have fine, wavy teeth; hairless or nearly so. Flowers bell-shaped, with 4 greenish-yellow sepals; ³⁄₁₆″ wide; in clusters at base of leaves; appearing in late spring. Fruit black and berrylike; ⁵⁄₁₆″ in diameter; pulp bitter; maturing in late summer or fall.

Habitat   Open woods and clearings with dry soils; planted along roadsides and fences.

Range   S. Ontario to Atlantic Coast, south to North Carolina, west to NE. Kansas and North Dakota; introduced from Europe and Asia.

Witch-hazel *Hamamelis virginiana*

According to legend, the forked branches of the Witch-hazel can be used as a divining rod to detect underground sources of water. An aromatic liquid extracted from the bark is used to make astringent lotions. The fruit capsule, when dried, contracts, and can shoot the seeds a distance of 30 feet.

Identification Height: 20–30′; diameter: 4–8″. Shrub or small tree with spreading branches forming a broad, open crown. Leaves dull dark green above, paler beneath; 3–5″ long and 2–3″ wide; broadly elliptical, with a pointed or rounded tip; unequal at base; broadening at the middle and becoming wavy-lobed. Flowers yellow, with 4 twisted, elongated petals; appearing on leafless twigs in late fall or winter. Fruit an irregular, brownish capsule, ½″ long; dividing into 2, and containing 1 or 2 dark, shiny seeds; maturing in fall.

Habitat In understory of hardwood forests with moist soil.

Range S. Ontario to Atlantic Coast, south to central Florida and west to Mississippi Valley; to 5000′; sometimes higher in southern Appalachians.

54

Apple *Malus sylvestris*

Many improved commercial varieties of apples are descended from this species, which has been in cultivation since ancient times. Introduced from Europe and western Asia, the Apple was successfully promoted in this country by "Johnny Appleseed"—whose real name was Jonathan Chapman.

Identification Height: 30–40′; diameter: 1–2′. Tree with a short trunk and rounded, spreading crown. Leaves green above, with dense gray hairs on undersurface; 2½–3″ long, 1¼–2¼″ wide; oval or elliptical, with sharp point at tip; edges have wavy sawteeth. Flowers white with pink tinge; 1¼″ wide; with 5 rounded petals; appearing in spring. Fruit a round, yellow or red apple; 2–3½″ in diameter; maturing in late summer or fall.

Habitat In clearings with moist soils; planted in gardens and along roadsides.

Range S. Canada and throughout much of the East; native to Europe and W. Asia.

Pear *Pyrus communis*

Like the Apple, this species is native to Europe and western Asia, and has been naturalized throughout much of North America. The Pear is prized for its wood, which is used to make fine furniture, drawing instruments, and rulers; it is exceptionally well suited to carving. There are several commercial cultivars, producing an abundant variety of delectable fruit.

Identification   Height: 40′; diameter: 1′. Slender tree with broad crown. Leaves shiny green above, paler below; 1½–3″ long and 1–2″ wide; broadly oval or elliptical, with very long stalk; edges with fine, wavy sawteeth. Flowers white; 1¼″ wide, with 5 rounded petals; appearing in clusters in early spring, with leaves. Fruit a green, yellow, or brown pear; 2½–4″ long; maturing in late summer.

Habitat   Clearings with moist soils; also planted along roads and in gardens and orchards.

Range   Maine to Florida, Texas, and Missouri; naturalized where it occurs; native to Europe and western Asia.

Flowering Dogwood *Cornus florida*

The beautiful blossoms of the Dogwood are a welcome sign of spring throughout the East. These trees are popular in yards and gardens; the wood, which is hard and resistant to shock, is used for a variety of small objects, such as pulleys and jewellers' blocks, as well as for loom shuttles.

Identification	Height: 30′; diameter: 8″. Small tree with a short trunk and lovely crown of spreading or nearly horizontal branches. Leaves green and smooth above, paler below, with fine hairs; 2½–5″ long and 1½–2½″ wide; opposite; elliptical with slightly wavy edges; 6–7 long, curved veins on each side; turning bright scarlet in fall. Blossoms white or pink, 1½–2″ wide, with 4 petal-like bracts; the actual flowers are tiny (³⁄₁₆″ wide) and crowded into a head ¾″ wide; appearing in spring, before leaves. Fruit berrylike, elliptical, bright red; ³⁄₈–⁵⁄₈″ long; in small cluster at end of long stalk; maturing in fall.
Habitat	Hardwood forests, uplands, and valleys.
Range	S. Ontario and Great Lakes states east to Maine and south to N. Florida and central Texas; to 4000′.

Black Cherry *Prunus serotina*

A familiar and widespread species, Black Cherry is found almost everywhere in the East. The wood of this species is used often in furniture and cabinetry, and the bark and fruit are used to make cough syrup, wine, and jelly. The bark may be helpful in identification, for it gives off a pungent aroma when crushed.

Identification Height: 80'; diameter: 2'. Tall tree with oblong or elliptical crown. Leaves shiny dark green above, lighter below, often with hairs along midvein; 2–5″ long and 1¼–5″ wide; elliptical, with 1 or 2 dark red glands at base; with finely saw-toothed edges; turning red or yellow in fall. Flowers white, in large clusters 4–6″ long; each blossom white, ⅜″ wide, with 5 rounded petals; appearing in late spring. Fruit an edible cherry; dark red, turning black; ⅜″ long; maturing in late summer.

Habitat Moist soils, except those that are very dry or very wet; sometimes in pure stands.

Range S. Quebec to Atlantic Coast, south to central Florida and west to E. Texas; absent from Mississippi delta region; to 5000' in southern Appalachians.

American Plum *Prunus americana*

This species is popular for its delicious edible fruit as well as for the large white flowers that appear in early spring. Because the species spreads by root sprouts, it is often useful in erosion control. Like many of our most popular fruit trees, including the Apple, Pear, and the cherries, the American Plum belongs to the rose family.

Identification   Height: 30′; diameter: 1′. Shrub or small tree with a broad crown formed of many spreading branches; often in thickets. Leaves dull green above, with veins slightly sunken; paler below with a few hairs; 2½–4″ long and 1¼–1¾″ wide; elliptical, sharply saw-toothed, and with a long point at tip. Flowers white, in clusters of 2–5; each flower ¾–1″ wide, with 5 rounded petals; odor slightly unpleasant; appearing in early spring before leaves. Fruit a red, round plum, ¾–1″ in diameter; edible; maturing in summer.

Habitat   Valleys and low slopes with moist soils.

Range   SE. Saskatchewan to New Hampshire; south to Florida, Oklahoma, and Montana; to 3000′ southward; higher locally.

64

Pin Cherry *Prunus pensylvanica*

The tiny, sour cherries of this species are used to make jelly and provide food for a variety of birds and animals. Also called Fire Cherry, it is one of the first trees to sprout in burned-over areas after a forest fire. Since the Pin Cherry does well on moist, sandy soils, it is common in areas covered by glaciers during the Ice Age.

Identification   Height: 30'; diameter: 1'. Small tree or shrub with narrow, rounded crown of horizontal branches. Leaves shiny green above, paler below; lance-shaped; 2½–4½" long and ¾–1¼" wide; long-pointed, with sharp sawteeth; turning bright yellow in fall. Flowers white, in clusters of 3–5; each flower ½" wide, with 5 rounded petals; appearing in spring with leaves. Fruit a small red cherry, ¼" in diameter; maturing in summer.

Habitat   Moist soils, especially clearings and burned-over areas; in pure stands or with aspens, Paper Birch, or Eastern White Pine.

Range   British Columbia to Newfoundland, south of Hudson Bay; south to Great Lakes region, New England, and Georgia; to 6000' in southern Appalachians.

Black Tupelo *Nyssa sylvatica*

This is a popular and attractive ornamental; its berrylike fruit is eaten by a variety of wildlife, and the flowers are also a good source of honey. The subspecies *biflora*, known as Swamp Tupelo, has narrower leaves; it occurs in swampy regions from Delaware to Texas. The Black Tupelo is also known as Blackgum and Pepperidge.

Identification Height: 50–100'; diameter: 2–3'. Tree with dense crown, either conical or flat; branches thin and nearly horizontal. Leaves shiny green above and paler below, often with hairs; 2–5" long and 1–3" wide; elliptical, usually without teeth; turning bright red in fall. Flowers greenish, appearing in spring with leaves; male flowers tiny, in heads ½" wide; female flowers each 3/16" wide, in clusters of 2–6; male and female usually on different trees. Fruit berrylike; blue-black, elliptical, 3/8–½" long; maturing in fall.

Habitat Hardwood and pine forests in valleys or uplands with moist soils.

Range Central Michigan to SW. Maine, south to S. Florida and E. Texas; to 4000'.

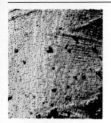

Hackberry *Celtis occidentalis*

This species is often a host to mites and fungi, whose activity deforms the branches into odd, bushy-looking growths called witches'-brooms. The wood of the Hackberry is used to make crates, plywood, and athletic equipment, and is marketed as soft elm. The Hackberry is a popular shade tree, especially in the South.

Identification Height: 50–90′; diameter: 1½–3′. Tree with spreading or drooping branches forming a rounded crown. Leaves in 2 rows along twigs; shiny green above, usually smooth; paler below, often with hairy veins; 2–5″ long, 1½–2½″ wide; oval, long-pointed, with sharp teeth; uneven at base; turning yellow in fall. Flowers greenish; ⅛″ wide; appearing at base of new leaves in early spring. Fruit berrylike; orange-red to dark purple-black; ¼–⅜″ in diameter; on a slender stalk, maturing in fall.

Habitat River valleys; also in mixed hardwood forests on upland slopes.

Range North Dakota east to New England, south to N. Georgia and NW. Oklahoma; local in S. Quebec and S. Manitoba; to 5000′.

70

Osage-orange *Maclura pomifera*

This spiny tree is a popular ornamental in the East. Its fruit is a distinctive greenish ball, known as a mock-orange, a horse-apple, or a hedge-apple. Native Americans used the wood for making bows, and early European settlers to this country used the root bark to make a yellow dye. French colonists called this tree *bois-d'arc* ("bow-tree"); that name has been corrupted to Bodark.

Identification Height: 50′; diameter: 2′. Medium-size tree, often with crooked trunk and irregular crown. Leaves shiny dark green above, paler below; 2½–5″ long and 1½–3″ wide; oval, with a long, pointed tip; lacking teeth; turning yellow before shedding in fall. Flowers tiny and greenish; many in round clusters ¾–1″ wide; appearing in early spring; male and female on different trees. Fruit a heavy, yellowish-green, fleshy, hard ball; 3½–5″ in diameter; maturing in fall.

Habitat River valleys with moist soils.

Range Native range uncertain; widely planted throughout much of the East.

72

Shagbark Hickory *Carya ovata*

This large tree is distinctive because of its shaggy bark, which separates from the trunk in narrow, curved strips. Native Americans once boiled the kernels to make a sweet hickory milk. The strong wood of the slow-growing hickories is excellent material for baseball bats; it is also used to smoke meats, imparting a distinctive flavor.

Identification Height: 70–100′; diameter: 2½′. Large tree with a tall trunk and irregular, narrow crown. Leaves yellowish green above, paler below; pinnately compound; 8–14″ long, usually with 5 elliptical leaflets, each 3–7″ long with fine sawteeth and hairy edges; turning yellow in fall. Flowers tiny, greenish; male in slender, compact catkins appearing in 3's; female flowers in groups of 2–5, on same twig; appearing in early spring before leaves. Fruit round, green; 1¼–2½″, with grooves; turning brown or black; elliptical nut within has edible seed.

Habitat Mixed hardwood forests in valleys or uplands with moist soils.

Range SE. Minnesota to SW. Maine, south to Georgia and SE. Texas; to 2000′ in North, higher in southern Appalachians.

74

Pignut Hickory *Carya glabra*

This species is widespread in the East and especially abundant in the southern Appalachians. The wood is used for skis and tool handles; it was formerly a popular material for wagon wheels. The similar Red Hickory (var. *odorata*) occurs in much the same range; it usually has seven leaflets, and its fruit is split to the base.

Identification — Height: 60–80′; diameter: 1–2′. Tall tree with spreading, irregular crown. Leaves light green, nearly stalkless; pinnately compound; 6–10″, usually with 5 leaflets, each 3–6″ long; lance-shaped with fine sawteeth; turning yellow in fall. Flowers tiny, greenish; male flowers in drooping, compact catkins, appearing in 3's; female flowers in groups of 2–10 at end of same twig; appearing in early spring before leaves. Fruit greenish; rounded or pear-shaped, 1–2″ long, with thin husk; turning dark brown and splitting to middle; thick-shelled nut within; maturing in late fall.

Habitat — Hardwood forests with oaks and other hickories.

Range — Illinois to S. New England; south to central Florida and extreme E. Texas; to 4800′ in southern Appalachians.

76

Bitternut Hickory *Carya cordiformis*

The twigs of this species have bright yellow buds, which make identification easy. The Bitternut is one of the most common hickories in the East; its nuts, unpalatable to humans, are likewise disdained by most animals. The Bitternut Hickory has been introduced to Europe, where it is cultivated in large gardens and botanical collections.

Identification Height: 60–80′; diameter: 1–2′. Tree with rounded, broad crown and tall trunk. Leaves yellowish green above, pale green below with fine hairs; pinnately compound, 6–10″ long with 7–9 leaflets, each 2–6″ long; lance-shaped, with fine sawteeth; turning yellow in fall. Flowers tiny, greenish; male in drooping, compact catkins, in 3's; female flowers in groups of 1–2 at tip of same twig; appearing in early spring, before leaves. Fruit round; thin husk has tiny yellowish scales; splitting into 4; smooth nut within has bitter seed.

Habitat Mixed hardwood forests; in moist valleys of the South, dry uplands northward.

Range Minnesota to SW. New Hampshire, south to NW. Florida and E. Texas; to 2000′.

White Ash *Fraxinus americana*

The White Ash has contributed greatly to sports in North America; its timber is used to make all kinds of athletic equipment, from baseball bats and polo mallets to tennis racquets and oars. This species is a popular ornamental and shade tree in the East, and it is cultivated in some parts of Europe.

Identification Height: 80'; diameter: 2'. Large tree with rounded or conical crown of dense foliage. Leaves dark green above, whitish below; opposite and pinnately compound, 8–12" long, usually with 7 leaflets, each 2½–5" long and 1¼–2½" wide; leaflets oval or elliptical, often almost lacking teeth; turning purple or yellow in fall. Flowers purplish, ¼" long; crowded in small clusters; male and female on different trees; appearing before leaves. Fruit a long, brownish key with a narrow wing; 1–2" long; in hanging clusters; maturing in late summer and fall.

Habitat Valleys and slopes with moist, well-drained, rich soil; often in forests with other hardwoods.

Range E. Minnesota to Cape Breton Island, south to N. Florida and E. Texas; to 2000' in the North, 5000' in the South.

80

Black Walnut *Juglans nigra*

The beautiful wood of the Black Walnut is especially prized in furniture-making, but because mature trees are scarce, the wood is extremely expensive. The edible walnuts are a delicious treat when they mature in fall. There are several cultivated varieties of this species that are the source of commercial walnuts.

Identification Height: 70–90'; diameter: 2–4'. Large tree with rounded, open crown and aromatic foliage. Leaves green to dark green, pinnately compound; 12–24" long, with 9–21 leaflets, each 2½–5" long, broadly lance-shaped with fine sawteeth; smooth above, with fine hairs below; turning yellow in fall. Flowers greenish, small; male in catkins, with 20–30 stamens; female 2-lobed, in groups of 2–5 at tip of same twig. Fruit roundish, with green husk and edible walnut; 1½–2½" wide; single or in pairs.

Habitat Well-drained, moist soils, often along streams, or in mixed forests.

Range SE. South Dakota to Connecticut and Massachusetts; south to NW. Florida, west to central Texas; absent from southern coasts; to 4000'.

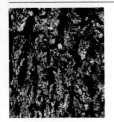

Pecan *Carya illinoensis*

Pecan wood is well suited to a variety of purposes, including floors, furniture, and veneers. Pecans are cultivated in the South for their edible nuts, and improved varieties have been developed. Pecans and hickories belong to the walnut family; their genus name, *Carya*, comes from the Greek for "walnut tree."

Identification Height: 100′; diameter: 3′. Tall tree with rounded, broad crown of large branches. Leaves yellowish green above, paler below; pinnately compound, 12–20″ long, with 11–17 leaflets, each 2–7″ long; slightly sickle-shaped, with fine sawteeth and short stalk; turning yellow in fall. Flowers tiny, greenish; male flowers with 5–6 stamens; in drooping catkins, often in 3's; female flowers in groups of 2–10 at tip of same twig; appearing in early spring before leaves. Fruit oblong, nutlike, with thin husk and edible seed; turning brown in fall.

Habitat Floodplains and valleys with loamy, well-drained, moist soils; in mixed hardwood forests.

Range W. Iowa and Indiana to Louisiana and S. Texas; to 1600′.

Boxelder *Acer negundo*

Like the related maples, the Boxelder has distinctive winged fruits, called keys or samaras. It is widely planted as an ornamental and shade tree. Also known as the Ashleaf Maple, Boxelder was once tapped like Sugar Maples, and its sweetish sap was used to produce sugar and syrup.

Identification | Height: 30–60'; diameter: 2½'. Small to medium-size tree with a rounded, broad crown and short trunk. Leaves light green above, with few hairs; paler below; opposite and pinnately compound, 6" long; 3–7 leaflets 2–4" long and 1–1½" wide; sometimes lobed; oval, with long point at tip and coarse sawteeth; turning yellow or red in fall. Flowers yellow-green, ³⁄₁₆" long, with 5 sepals; clustered on drooping stalks; male and female on separate trees; appearing before leaves. Fruit a yellow key, 1–1½" long; forked and in pairs, with long, curved wing and 1 seed; maturing in summer and persisting.

Habitat | Stream banks and valleys with wet or moist soils.

Range | S. Alberta through Ontario and New York, south to S. Texas and central Florida; to 8000' in the Southwest.

Red-osier Dogwood *Cornus stolonifera*

This shrub is widespread in moist areas throughout the northern coniferous forests of North America. Its berries provide food for a variety of animals and birds. Red-osier Dogwood forms thickets where it grows; the branches often take root in the ground and form new shoots around an existing plant.

Identification Height: 3–10′, occasionally higher; diameter: 3″. Large, spreading shrub with several stems; rarely a small tree. Leaves opposite; dull green above, whitish below, with fine hairs; elliptical, with untoothed edges, 1½–3½″ long and ⅝–2″ wide, with 10–14 long, curved veins; turning red in fall. Flowers white, with 4 petals; ¼″ wide; in upright clusters 1¼–2″ wide; appearing in late spring or early summer. Fruit a white berry, ¼–⅜″ wide; maturing in late summer.

Habitat Areas with moist soils; often along streams or in forest understories.

Range Central Alaska and British Columbia to Newfoundland, south to N. Virginia and west to California; to 5000′.

American Elder *Sambucus canadensis*

The berries of this large shrub are a favorite for making pies, jellies, and preserves; elderberry wine is a popular home-brewed beverage, made famous in the play *Arsenic and Old Lace*. American Elder is widespread and common throughout much of the East; it sprouts from roots, and is often found near water.

Identification Height: 16′; diameter: 6″. Large shrub or small tree with a few stout branches forming an irregular crown. Leaves shiny green above, duller below, with hairs along midvein; opposite and pinnately compound, 5–9″ long, with 3–7 leaflets, each 1½–4″ long and ¾–2″ wide; leaflets elliptical with sharp sawteeth. Flowers white, with 4 or 5 lobes; ¼″ wide; in fragrant clusters 4–8″ wide; appearing in late spring or early summer. Fruit a black or purple berry; ¼″ in diameter; maturing in late summer or fall.

Habitat Open areas with wet soils, especially near water at forest edges.

Range SE. Manitoba to Nova Scotia, south to Texas and Florida; to 5000′.

Smooth Sumac *Rhus glabra*

This sumac is the only tree native to all 48 contiguous states. Many kinds of birds and small mammals consume the fruits, which can also be used to make a kind of drink; Native Americans used young sprouts as salad. Contact with the sap of the related Poison-sumac (*Toxicodendron vernix*) causes a severe rash.

Identification   Height: 20′; diameter: 4″. Large shrub or small tree with stout branches forming a flat, open crown. Leaves shiny green above, whitish below; pinnately compound, 12″ long with 11–31 lance-shaped, saw-toothed, smooth leaflets, each 2–4″ long; turning reddish in fall. Flowers small, white; less than ⅛″ wide, with 5 petals; in large, upright clusters 8″ long; male and female on separate trees; appearing in early summer. Fruit dark red, berrylike, ⅛″ or wider with short, sticky red hairs; in clusters; maturing in late summer and persisting.

Habitat   Forest edges, open uplands, clearings, and roadsides; often in sandy soils.

Range   E. Saskatchewan to Maine, south to central Texas and NW. Florida; to 4500′ in the East.

Ailanthus *Ailanthus altissima*

This sturdy tree has shown remarkable tolerance for conditions in developed areas, and it is widely planted in cities. It makes a good ornamental or shade tree, but the roots are believed to be toxic, and they often find a way to penetrate into wells and springs. Crushed leaves and the male flowers of this species have an unpleasant odor. Also called Tree-of-Heaven.

Identification Height: 50–80′; diameter: 1–2′. Medium-size to tall tree with rounded, open crown and thick branches. Leaves green above, paler below; pinnately compound, 12–14″ long, with 13–25 leaflets, each 3–5″ long and 1–2″ wide; lance-shaped, with 2–5 teeth at base. Flowers yellow-green, ¼″ long, with 5 petals; in clusters; male and female usually on different trees; appearing in late spring or early summer. Fruit a key, 1½″ long; red-green or reddish brown; narrow, with 1 seed; maturing in late summer and fall.

Habitat Open areas, old fields, roadsides, and cities.

Range Widely naturalized across North America; native to China.

Weeping Willow *Salix babylonica*

The bright green leaves of the Weeping Willow are among the earliest to appear in spring and often the last to disappear in fall. This familiar willow can easily be identified by its distinctive, drooping branches and foliage; they give it a "weeping" silhouette. This tree was long believed to be the tree referred to in Psalm 137: "By the rivers of Babylon, there we sat down, yea, we wept, when we remembered Zion. We hanged our harps upon the willows in the midst thereof."

Identification   Height: 30–40'; diameter: 2', sometimes larger. Small to medium-size tree with short trunk and drooping branches. Leaves dark green above, gray or white below; 2½–5" long, narrowly lance-shaped with pointed tips; edges with fine sawteeth. Flowers greenish, in catkins ⅜–1" long; appearing in early spring. Fruit a light brown capsule, 1/16" long; maturing in late spring or early summer.

Habitat   Open areas, parks, and gardens; often near water.

Range   S. Ontario and Quebec to Missouri and Georgia; native to China.

Honeylocust *Gleditsia triacanthos*

This species is easy to identify because the trunk is covered with large clumps of stout, branched spines. The Honeylocust is a popular shade tree. It produces long brown pods that are relished by animals. Today there are only 11 species of *Gleditsia*, relics of a genus that flourished 70 million years ago.

Identification   Height: 80′; diameter: 2½′. Large tree with open crown and clumps of brown spines on trunk. Leaves shiny dark green above, paler and smooth below; pinnately compound, 4–8″ long, with many wavy-edged leaflets, each ⅜–1¼″ long; turning yellow in fall. Flowers yellow-green, ⅜″ long, bell-shaped, hairy, with 5 petals; male and female flowers often on different trees; appearing in clusters in late spring. Fruit a flat, dark brown pod, 6–16″ long and 1¼″ wide, containing many beanlike seeds; falling unopened in late autumn or winter.

Habitat   Floodplains and areas with moist soils in mixed forests; sometimes in drier areas upland.

Range   South Dakota to Pennsylvania, south through most of Mississippi Valley and in Appalachians; to 2000′.

Black Locust *Robinia pseudoacacia*

A popular shade tree, the Black Locust is useful in erosion control and is often planted on lands that have been reclaimed after strip-mining. It was perhaps one of the earliest native trees to be discovered by European colonists; today it is cultivated for ornament in many parts of Europe and is naturalized in some places.

Identification Height: 4–80'; diameter: 1–2'. Medium-size tree with irregular crown and forking trunk, often crooked. Leaves dark blue-green above, pale and smooth below; pinnately compound; 6–12" long with 7–19 leaflets, each 1–1¾" long and ½–¾" wide; elliptical, with tiny bristle at tip; edges smooth; hairy when young, becoming smooth. Flowers white, ¾" long, and pealike, with 5 irregular petals; in clusters 4–8" long; fragrant; appearing in late spring. Fruit a narrow, brownish pod, 2–4" long, with several brown, beanlike seeds; maturing in fall and persisting in winter.

Habitat Open areas with sandy soils; also in woodlands.

Range Pennsylvania and Ohio to Alabama and Georgia; also from S. Missouri to E. Oklahoma; 500' to 5000'.

100

Southern Catalpa *Catalpa bignonioides*

This tree's original range was probably quite small, limited to the southernmost area of the Mississippi Valley north of the delta. But it is a popular ornamental and shade tree, with abundant multicolored flowers, and today it has been naturalized throughout much of the eastern United States.

Identification
Height: 50′; diameter: 2′. Tree with short trunk and spreading branches forming a rounded crown. Leaves dull green above, paler below with soft hairs; heart-shaped, 5–10″ long and 4–7″ wide at broadest point, with abrupt long point at tip and notch at base; edges smooth; turning blackish in fall. Flowers white, with orange stripes and purple spots; bell-shaped, 1¼″ wide, with 5 fringed lobes; in upright, branched clusters to 10″ long; in late spring. Fruit a cigarlike brown capsule, 6–12″ long and ⁵⁄₁₆–³⁄₈″ wide; dividing into 2 parts; maturing in fall.

Habitat
Roadsides and clearings, and other open areas with moist soils.

Range
Widely naturalized from Michigan to S. New England and south to Texas and Florida; 100–500′.

102

Paper Birch *Betula papyrifera*

These beautiful tall trees often form pure stands in the North. The bark is smooth, thin, and papery; Native Americans used it to make lightweight canoes, stretching the material over a frame of Northern White-cedar. The wood is used to make a variety of small objects, such as toothpicks and spools.

Identification   Height: 50–70'; diameter: 1–2'. Tree with narrow crown of horizontal or slightly drooping branches. Leaves dark green above, pale yellow-green below; oval, long-pointed, with coarse double sawteeth and usually with many veins; turning light yellow in fall. Flowers tiny; male flowers yellowish, in drooping catkins; female flowers greenish, in short, erect catkins; appearing in early spring. Cone narrow, cylindrical, brownish, 1½–2" long, on slender stalk; maturing in fall.

Habitat   Upland areas with moist soils.

Range   NW. Alaska to Labrador, south to Montana and Oregon in the West, to New York and New England in the East; to 4000'.

104

White Mulberry *Morus alba*

The leaves of this species are the main food of silkworms; because silk has long been important to the economy of many Asian nations, the White Mulberry has been cultivated since time immemorial. The abundant berries are consumed in large quantities by birds and mammals as well as by people.

Identification   Height: 40′; diameter: 1′. Small tree with rounded crown. Leaves shiny green above, paler with some hairs below; in 2 rows; 2½–7″ long and 2–5″ wide; oval with pointed tip or deep, irregular lobes; with coarse sawteeth and 3 veins extending from base. Flowers greenish, tiny, in short, crowded clusters; male and female often on different trees. Fruit raspberrylike; usually white, sometimes purplish pink; ⅜–¾″ long; appearing in late spring.

Habitat   Dry, warm areas; tolerant of cities and developed areas.

Range   Naturalized widely throughout the East and cultivated throughout; native to China.

106

Gray Birch *Betula populifolia*

In the Northeast, the Gray Birch is a familiar sight on burned-over areas and clearings. It is a pioneer species, often growing in abandoned areas and protecting the seedlings of forest trees that will eventually outlive it. The smooth white bark of this tree—which is also known as White Birch and Wire Birch—is thin but not papery.

Identification Height: 30′; diameter: 1′. Small tree with conical or pointed crown and thin, short branches; often in groups all growing from base of an old trunk. Leaves shiny dark green above, paler below, with hairs along midvein; triangular, with wide base and narrow tip; 2–3″ long and 1½–2½″ wide; with sharp double sawteeth on edges and 8–16 veins; turning pale yellow in fall. Flowers tiny; male yellowish, in drooping catkins; female greenish in erect catkins on same twig. Cones brownish, cylindrical, ¾–1¼″ long, with many hairy scales; maturing in fall.

Habitat Abandoned uplands and open areas; also in mixed woodlands with moist soils.

Range S. Ontario to Atlantic Coast, south to New Jersey and Pennsylvania; local farther south; to 2000′.

108

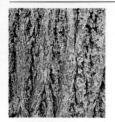

American Basswood *Tilia americana*

This tree is the northernmost representative of the basswoods; there are three native species, all important trees of the eastern forests and all favorite shade trees. Basswoods are also called linden trees; close relatives are found throughout Europe, where their popularity dates from the days of ancient Greece and Rome.

Identification   Height: 60'; diameter: 2–3'. Large tree with small, sometimes drooping branches forming a dense crown; often with 2 or more trunks. Leaves shiny dark green above, lighter below with a few hairs in tufts; in 2 rows; 3–6" long and nearly as wide; rounded or heart-shaped with long-pointed tip and coarse sawteeth along edges; pale yellow or brown in fall. Flowers yellowish white; ½–⅝" long, with 5 petals; in fragrant clusters; early summer. Fruit nutlike, rounded, hard; ⅜" in diameter; gray with fine hairs; maturing in late summer or fall.

Habitat   Uplands and valleys with moist soils; also in hardwood forests.

Range   SE. Manitoba to Maine, south to NE. Oklahoma and W. North Carolina; to 3200'.

Lombardy Poplar *Populus nigra*

The Lombardy Poplar grows tall and narrow, like a green column; throughout much of the East it is a familiar sight along roads and in fields, where it is planted in long rows and used as a windbreak or shelterbelt. It is not a species, but actually a cultivated variety of a wild European species of the same name; it must be propagated from cuttings, as all the trees are male and never bear seeds.

Identification   Height: 30–60′; diameter: 1–2′. Medium-size to tall tree with straight trunk and close crown of upright branches. Leaves green above, paler below; triangular, 1½–3″ long and as wide at base; edges with wavy sawteeth. Flowers tiny; male flowers in catkins 2″ long; appearing in early spring before leaves. No female flowers present; no fruit.

Habitat   Temperate areas with moist soils; along roads, in cultivated fields, and in gardens.

Range   Planted throughout most of United States; native to Europe and W. Asia.

Quaking Aspen *Populus tremuloides*

The most widely distributed tree in North America, the Quaking Aspen never seems to be still. Each leaf grows on a long, flattened stalk, and the presence of even the gentlest breeze will set the leaves in motion. In France, aspens are known as *"langue de femme";* in Wales, they are called *"coed tafod merched."* Both names mean "woman's tongue," apparently an allusion to the incessant movement of the leaves.

Identification   Height: 40–70'; diameter: 1–1½'. Medium-size to tall tree with narrow, rounded crown of rather sparse foliage. Leaves shiny green above, duller below, nearly round, with short-pointed tip; 1½–3" long and about as wide; turning gold in fall. Flowers tiny, brownish, in catkins 1½–2" long; male and female on different trees; appearing in early spring, before leaves. Fruit a narrow, green, conical capsule, ¼" long, in drooping clusters to 4" long; maturing in late spring.

Habitat   Gravelly slopes; moist soils; often in pure stands.

Range   Manitoba to S. Wisconsin, east to Atlantic Coast; to sea level in the North; 6500–10,000' in the South.

114

Ginkgo *Ginkgo biloba*

The Ginkgo is a survivor from prehistoric times; despite its leafy appearance, its closest relatives are conifers. This durable introduced species is familiar in eastern North America, where it thrives along city streets, in parks, and on lawns. The female tree bears a foul-smelling pulpy seed; the kernel is edible, and apparently tasty, for it is considered a delicacy in the Orient.

Identification   Height: 50–70′; diameter: 2′. Tree with straight trunk and open crown, pointed on young trees and spreading with age. Leaves dull light green, fan-shaped with narrow base and broad, wavy end; 1–2″ long and 1½–3″ wide; often 2-lobed; turning yellow in fall. Flowers (rarely occurring) greenish; male flowers in catkins ¾″ long; female flowers in small, acornlike catkins; on different trees. Female fruit (technically a pulp-coated yellowish seed), 1″ long, with one kernel; maturing and shedding in fall; male or pollen cone ¾″ long.

Habitat   Temperate areas with high humidity and moist soils.

Range   Planted widely throughout the East; probably native to SE. China.

116

Yellow-poplar *Liriodendron tulipifera*

The Yellow-poplar—familiarly known as the Tuliptree or Tulip-poplar—is one of the most beautiful hardwoods of the East; Walt Whitman called it the "Apollo of the woods," in reference to its tall, graceful, columnar trunk. In spring, this tree produces striking bright orange and green, tuliplike flowers.

Identification Height: 80–120′; diameter: 2–3′. Very tall tree with long, straight trunk and a narrow crown, becoming broader with age. Leaves shiny dark green above, paler below; 3–6″ long and as wide; saddle-shaped, with long straight base and 4 (rarely 6) distinct, paired lobes; on a long stalk. Flowers large, cup-shaped; 6 green petals with orange base; upright and single, appearing in spring. Fruit conelike, light brown, with many overlapping nutlets; 2½–3″ long; maturing in fall and shedding.

Habitat Valleys and slopes with moist, well-drained soils; often in pure stands.

Range Michigan and S. Ontario to Vermont, south to Louisiana and Florida; to 1000′ in the North; to 4500′ southward.

118

Sassafras *Sassafras albidum*

A small tree or thicket-forming shrub, Sassafras is common and widespread in eastern North America; its unusual, mitten-shaped leaves and the sweet, root beer-like aroma of the roots and bark make it easy to identify. The common name is believed to be a corruption of the Spanish *saxafrax* (saxifrage).

Identification   Height: 30–60'; diameter: 1½'. Small to medium-size tree or shrub with variable leaves and narrow, spreading crown. Leaves shiny green above, paler and sometimes hairy below; elliptical; simple or 2- or 3-lobed, all on same stalk; 3–5" long and 1½–4" wide; turning orange, red, or yellow in fall. Flowers yellow-green, ⅜" long, in clusters; male and female usually on different trees; appearing in early spring before leaves. Fruit a blue-black berry; elliptical, ⅜" long; in a reddish cup with red stalk; maturing in fall.

Habitat   Moist uplands, valleys, old fields, and forest clearings, often in sandy soils.

Range   Central Michigan and S. Ontario to SW. Maine, south to central Florida and E. Texas; to 5000' southward.

120

White Poplar *Populus alba*

Like its close relative the Quaking Aspen, the White Poplar has long-stalked leaves that tremble in the slightest breeze. This species is a native of Europe and Asia, but it was introduced to North America in colonial times and so is widespread today. It is a popular shade and street tree.

Identification   Height: 80′; diameter: 2′. Large tree with many branches, looking randomly placed on trunk, and irregular crown. Leaves dark green above, white and covered with dense hairs below; maplelike, oval, with 3 or 5 lobes, blunt tip, and small teeth; turning dull red in fall. Flowers tiny, in catkins 1½–3″ long, with dense white hairs; male and female on different trees; appearing before leaves in early spring. Fruit an egg-shaped capsule, ³⁄₁₆″ long; in drooping clusters with many tiny, cottony seeds; maturing in summer.

Habitat   Areas with moist soils; often along roadsides and the edges of cultivated fields.

Range   Widely naturalized throughout North America; native to Europe and Asia.

122

Eastern Cottonwood *Populus deltoides*

Another relative of the aspens, the Eastern Cottonwood is one of the largest hardwoods in the East. It is a popular shade tree and shelterbelt species, perhaps because it grows extremely fast; in some parts of its range, this cottonwood grows five feet a year, and may reach 13 feet its first year.

Identification Height: 100′; diameter: 3–4′. Very tall tree with large trunk and many stout branches; crown open, spreading. Leaves shiny green, triangular, with flat base and long-pointed tip; 3–7″ long and 3–5″ wide; with curved, coarse teeth; turning yellow in fall. Flowers tiny, brownish, in catkins 2–3½″ long; male and female on different trees; appearing in early spring. Fruit a light brown, elliptical capsule, ⅜″ long; maturing in spring and dividing; in clusters to 8″ long, with many tiny cottony seeds.

Habitat In wet soils, especially along streams; often in pure stands or with willows.

Range S. Alberta to S. Quebec and New Hampshire, south to NW. Florida and W. Texas; to 1000′ in the East.

Sycamore *Platanus occidentalis*

The Sycamore has greenish-brown bark that peels in rectangular strips to reveal a pale inner layer beneath—an easy field mark to recognize. Also called the American Planetree, this species develops the largest trunk of any North American hardwood; the current champion has a diameter of 11 feet.

Identification   Height: 60–100'; diameter: 2–4'; sometimes much larger. Tall tree with straight trunk rising from large base; broad, open crown of large, spreading branches. Leaves bright green above, paler below, with hairs only on veins; broadly oval, with squarish base and 3 or 5 lobes; wavy edges with a few teeth giving a scalloped look; turning brown in fall. Flowers tiny, greenish, in drooping, spherical clusters; male and female on separate twigs; appearing in spring. Fruit a nubbly brown ball, 1" in diameter, hanging on long stalk; maturing in fall.

Habitat   Stream banks, floodplains, and in wet areas at the edges of lakes and swamps; also in mixed forests and in parks.

Range   E. Nebraska to SW. Maine, south to central Texas and NW. Florida; to 3200'.

126

Red Maple *Acer rubrum*

In very early spring, distinctive reddish flower clusters appear along the still-leafless twigs of the Red Maple, brightening the winter landscape. This species also produces reddish fruit and leafstalks. But the most brilliant display of color comes in fall, when the leaves turn a fiery scarlet.

Identification	Height: 60–90′, sometimes taller; diameter: 2½′. Large tree with compact crown and red flowers, fruit, and leafstalks. Leaves dull green above, whitish and hairy below; opposite and broadly oval, with 3 shallow lobes; 2½–4″ long and about as wide; turning brilliant red, orange, or yellow in fall. Flowers reddish, ⅛″ long, in crowded clusters; male and female often on different trees; appearing in late winter or very early spring. Fruit a pair of reddish-brown keys or samaras; ¾–1″ long; maturing in spring.
Habitat	Stream banks, valleys, swamps, and other areas with moist soils; also in mixed hardwood forests.
Range	SE. Manitoba and N. Minnesota to Newfoundland, south to E. Texas, S. Florida; to 6000′.

Sweetgum *Liquidambar styraciflua*

The aromatic resin of the Sweetgum has been used to make various kinds of medicines as well as to manufacture chewing gum; the durable wood is of great importance in the furniture industry. Sweetgum leaves look like maple leaves but are alternate on the stem, instead of in pairs.

Identification Height: 60–100'; diameter: 1½–3". Large, straight, symmetrical-looking tree; conical crown becomes rounded with age. Leaves shiny dark green above, paler below; maplelike, with 5 or 7 fine-pointed lobes and saw-toothed edges; base notched; turning reddish in fall. Flowers tiny, greenish; male flowers in several drooping clusters; female flowers in a ball. Fruit a long-stalked brown ball, 1–1¼" in diameter; made up of many individual fruits, with prickly points on the surface; maturing in fall and persisting through winter.

Habitat Valleys, lower slopes, and mixed woodlands with moist soils.

Range SW. Connecticut to central Florida, west to S. Illinois and E. Texas; to 3000'.

Sugar Maple *Acer saccharum*

Early spring is the season for tapping Sugar Maples; the sweet sap of these trees flows abundantly then, and is used to make delicious maple sugar candies as well as syrup for pancakes and waffles. The leaf of the Sugar Maple is the familiar emblem on the national flag of Canada.

Identification   Height: 70–100′; diameter: 2–3′. Large tree with rounded crown of dense foliage. Leaves dull, dark green above, paler and sometimes hairy below; opposite; 3½–5½″ long and as wide, with 5 long, sharp-pointed lobes and a few jagged teeth along edges; fall foliage deep red, orange, and yellow. Flowers yellowish green, tiny, in bell-shaped, 5-lobed calyx; in hanging clusters on hairy stalk; appearing in spring with leaves. Fruit a pair of brownish keys, 1–1¼″ long; maturing in fall.

Habitat   Uplands and valleys with moist soils; sometimes in pure stands.

Range   SE. Manitoba to Nova Scotia, south to E. Kansas and North Carolina; local farther south; to 2500′ in the North; 3000–5000′ in southern Appalachians.

132

Norway Maple *Acer platanoides*

Introduced from Europe, the Norway Maple is a fast-growing, popular ornamental. It does well in cities and suburbs because it is tolerant of dust and soot. Its leaves are similar in shape to those of the Sycamore, but the two species are not related; the Norway Maple's leaves are always opposite, and those of the Sycamore are always alternate.

Identification   Height: 60'; diameter: 2'. Medium-size or tall tree with rounded, dense crown. Leaves dull green above, paler below; opposite; 4–7" long and wide with 5 lobes; with scattered teeth along edges and notched base; turning bright yellow in fall. Flowers greenish yellow; 5/16" wide, with 5 petals; in clusters; male and female usually on different trees; appearing in early spring before leaves. Fruit a pair of keys with widely spread wings; light brown; maturing in summer.

Habitat   Streets, roadsides, and gardens.

Range   Planted throughout much of United States; native to Europe.

134

Silver Maple *Acer saccharinum*

Like its relative the Sugar Maple, the Silver Maple also produces a sweet sap, but usually in small quantities. Silver Maple leaves have much deeper lobes than do the leaves of other maples. This feature and the distinctive bark, which becomes shaggy with age, make the Silver Maple an easy tree to recognize.

Identification	Height: 50–80′; diameter: 3′. Large tree with short trunk and long, curving branches forming an irregular crown. Leaves dull green above, silvery white below; opposite and broadly oval, 4–6″ long and about as wide; deeply lobed with long sawteeth; turning pale yellow in fall. Flowers reddish, turning greenish yellow; ¼″ long, in crowded clusters; appearing in late winter or very early spring, before leaves. Fruit a pair of widely forking keys; light brown, 1½–2½″ long; maturing in spring.
Habitat	Stream banks, floodplains, swamps, and other areas with wet soils; often with other hardwoods.
Range	N. Minnesota and S. Ontario to New Brunswick; south to E. Oklahoma and NW. Florida; to 2000′, sometimes higher.

136

Pin Oak *Quercus palustris*

This species is named for the long bristles, or "pins," on its leaves, at the tip of each pointed lobe. A very popular shade tree, it has an almost oval silhouette: The lowest branches droop somewhat, the middle ones grow horizontally, and the top ones point skyward. The Pin Oak grows in wet areas without deep duff; its small acorns do not have to penetrate leaf litter to take root, as the larger acorns of some other oaks must.

Identification Height: 50–90'; diameter: 1–1½'. Fairly tall tree with conical crown becoming less regular with maturity. Leaves shiny dark green above, paler and slightly shiny below; elliptical, 3–5″ long and 2–4″ wide, with 5–7 deep lobes; turning red or brown in fall. Acorns small, green, turning brown; nearly round, ½″ in diameter; one-fourth to one-third enclosed by thin, saucer-shaped cup; maturing in second year.

Habitat Wet, poorly drained areas, sometimes with clay soils; often in nearly pure stands; also parks and gardens.

Range Vermont and Massachusetts south to North Carolina and west to S. Iowa, Kansas, and NE. Oklahoma; to 1000'.

138

Black Oak *Quercus velutina*

The distinctive yellow or orange inner bark of this species gives rise to the alternate name Yellow Oak. This inner bark was once a source of tannin and of a yellow dye. Like the Pin Oak, the Black Oak has sharp bristles on the leaf tips. This species is also sometimes called Quercitron Oak.

Identification Height: 50–80′; diameter: 1–2½′. Medium-size to tall tree with open, narrow crown. Leaves shiny green above, yellowish green below; elliptical, 4–9″ long and 3–6″ wide, with 7–9 shallow or deep, sharp-pointed lobes; turning dull red or brown in fall. Acorns light brown or greenish brown; ⅝–¾″ long, elliptical, half enclosed by deep, top-shaped cup with border of rust-colored, hairy scales; maturing in second year.

Habitat Sandy and rocky ridges, dry uplands, and clay soils on hillsides; sometimes in pure stands.

Range SE. Minnesota and S. Ontario to SW. Maine, south to E. Texas and NW. Florida; to 5000′.

Northern Red Oak *Quercus rubra*

The Northern Red Oak is a handsome and popular ornamental; it is fast-growing and tolerates a variety of difficult conditions, from soot and city dirt to cold temperatures. This is an important timber species, used in making furniture and floors as well as pulpwood, fence posts, and pilings.

Identification Height: 60–90′; diameter: 1–1½′. Tall tree with stout, spreading branches forming a rounded crown. Leaves dull green above, paler below, with some hairs; elliptical; 4–9″ long and 3–6″ wide with 7–11 shallow, wavy lobes; a few bristles at tips. Acorn reddish- or brownish-green, egg-shaped; ⅝–1⅛″ long; less than one-third enclosed in reddish-brown cup of blunt scales; maturing in second year.

Habitat In a variety of soil conditions, from moist and loamy areas to rocky and even clay soils; often in pure stands.

Range Minnesota to Cape Breton Island; south to North Carolina and in mountains to Georgia; west to E. Oklahoma; to 5500′ southward.

142

White Oak *Quercus alba*

This is one of the largest oaks in North America, and it produces some of the finest timber on the continent. The wood is used in furniture-making and shipbuilding; it is also well suited to making barrels for whisky and sherry, giving rise to the White Oak's alternate common name, Stave Oak.

Identification Height: 80–100′; diameter: 3–4′. Tall tree with rounded crown formed of spreading, stout branches. Leaves bright green above, paler gray-green or whitish below; elliptical, 3–4″ long and 2–4″ wide, with 5–9 lobes; widest beyond middle and tapered at base; turning red or brown in fall. Acorns greenish, turning light gray; egg-shaped, ⅜–1¼″ long, one-fourth enclosed by shallow, bumpy cup; maturing in first year.

Habitat Uplands and lowlands with moist, well-drained soils; often in pure stands.

Range Central Minnesota and S. Ontario to Maine, south to E. Texas and N. Florida; to 5500′; higher in southern Appalachians.

144

Bur Oak *Quercus macrocarpa*

This species is easy to recognize by its distinctive acorns, which are almost entirely enclosed in a fringed, burlike cup—and which give the tree its common name. The Bur Oak's range is more northerly than that of any other oak in North America. A popular ornamental and shade tree, it is also called the Mossycup Oak.

Identification   Height: 50–80′; diameter: 2–4′. Medium-size to tall tree (sometimes a shrub) with broad, rounded crown of spreading branches. Leaves dark green above, paler and with fine hairs below; narrowly oval, with broad, rounded tip and tapered base; lower half with 2–3 deep lobes on each side; upper half with 5–7 shallow lobes on each side. Acorns yellowish green, rounded or elliptical; ¾–2″ long and as wide; almost entirely enclosed in fringed, burlike cup; maturing in first year.

Habitat   A variety of soils: gravelly ridges, sandy floodplains, loamy areas, and stream banks; often in pure stands.

Range   SE. Saskatchewan and North Dakota to New Brunswick, south to Tennessee, west to SE. Texas; local in Louisiana and Alabama; 300–2000′; higher toward northwest.

146

Horsechestnut *Aesculus hippocastanum*

A European species, the Horsechestnut arrived in Pennsylvania in 1746, where the noted naturalist John Bartram was the first citizen of the New World to cultivate it. The species evidently caught on in popularity, and today it is a favorite ornamental throughout the United States.

Identification Height: 70′; diameter: 2′. Fairly tall tree with stout branches forming a spreading rounded or elliptical crown. Leaves green or gray-green above, paler below; palmately compound, with 5–7 leaflets spreading like the fingers of a hand; each leaflet broadest toward tip, tapered at base; 4–10″ long and 1–3½″ wide, with saw-toothed edges. Flowers white, bell-shaped; 1″ long, with 4–5 petals and red or yellow spots at center; many in long, upright clusters, 10″ long; in late spring. Fruit a brown, spiny capsule, dividing into 2 or 3 parts; with 2 poisonous, chestnutlike seeds; in late summer.

Habitat Rich, moist soils; along streets or roads, and in gardens.

Range Widely planted throughout; naturalized in some parts of the Northeast; native to SE. Europe.

148

Ohio Buckeye *Aesculus glabra*

A close relative of the Horsechestnut, this species is the state tree of Ohio and is sometimes used as an ornamental. It has poisonous seeds, leaves, and bark. The wood is used for furniture and musical instruments, among other things. The name "buckeye" probably comes from the seed, which looks like the eye of a deer.

Identification Height: 30–70′; diameter: 1–2′. Medium-size (sometimes shrubby) with rounded, irregular crown. Leaves green above, paler below and often hairy; opposite, palmately compound, with 5–7 elliptical leaflets, each 2½–6″ long and ¾–2¼″ wide, with toothed edges; yellow in fall. Flowers yellow-green with an unpleasant odor; bell-shaped, ¾–1″ long, with 4 petals and 7 longer stamens; in upright clusters to 6″; in spring. Fruit a spiny brown capsule, 1–2″ in diameter, with 1–3 poisonous dark brown seeds; maturing in summer or fall.

Habitat Valleys and mountain slopes with moist, rich soils; may form thickets near streams; in mixed hardwood forests.

Range Central Iowa to W. Pennsylvania, south to SE. Oklahoma and central Alabama; 500–2000′.

150

Cabbage Palmetto *Sabal palmetto*

This is the familiar "palm tree" of Florida and the southern coastal areas of Georgia and the Carolinas. A decorative ornamental, this species also produces valuable timber that is used for docks and wharf pilings. The hearts of young leaf shoots are often used in salads, and are known as hearts of palm.

Identification   Height: 30–50', sometimes taller; diameter: 1½'. Medium-size evergreen palm with large thatch of fan-shaped leaves around top. Leaves shiny dark green; 4–7' long and about as wide, folded into long, narrow segments; coarse and stiff, often drooping. Bark partly covered near crown by aged brown leafstalks. Flowers whitish, ³⁄₁₆" long, and fragrant; appearing in drooping or curved clusters in early summer. Fruit a shiny black berry, ³⁄₈" in diameter; maturing in fall.

Habitat   Coastal areas; inland in hammocks on sandy soil.

Range   SE. North Carolina south along coast of Florida, including Keys; also inland in Florida.

Baldcypress *Taxodium distichum*

Rising up out of the muddy water of swamps, the Baldcypress is a striking and characteristic species of the wetlands of the Southeast. It often has conical "knees" that project up out of the water from submerged roots. A variety called Pondcypress (var. *nutans*) grows in very wet areas of the South.

Identification Height: 100–120′, sometimes taller; diameter: 3–5′, sometimes wider. Large tree with deciduous crown of needles and wide, flat crown of spreading branches; trunk often enlarged or buttressed at base. Needles light green above, paler below, ⅜–¾″ long; single, flat, in crowded, featherlike rows; turning brown and shedding in fall. Cones gray, round, ¾–1″ in diameter; single or in pairs at end of twig; shedding scales in fall. Pollen cones tiny, in long, narrow clusters to 4″.

Habitat Swamps and very wet soils along riverbanks and in floodplains; often in pure stands.

Range S. Delaware along coast to S. Florida; Mississippi Valley from SW. Indiana to Louisiana and parts of S. Texas; to 500′; to 1700′ locally in Texas.

154

Eastern Redcedar *Juniperus virginiana*

The fragrant, red-grained wood of cedar closets and chests comes from this familiar and widespread conifer. The Eastern Redcedar is well adapted to living in a variety of conditions, as it tolerates temperature and moisture extremes better than any other North American conifer. The cones—which look like juicy berries—are a staple in the diet of many kinds of wildlife, and the tree is a popular ornamental and shelterbelt species in much of the eastern half of the United States.

Identification   Height: 40–60'; diameter: 1–2'. Medium-size evergreen with narrow, straight, dense crown. Leaves dark green; tiny (1/16") and scalelike; opposite, in 4 rows along angled twigs. Berrylike female "fruit"—technically a cone—dark blue with whitish bloom; male pollen cones on separate trees.

Habitat   Limestone uplands and other dry areas, floodplains and wetlands, and abandoned fields; often in pure stands.

Range   North Dakota to Maine, south to Texas and N. Florida; to 3000' in the South.

Northern White-cedar *Thuja occidentalis*

This species is also known as the Eastern Arborvitae. It was one of the first trees to be discovered by European settlers to this country, and a tea made from the bark and leaves is credited with having saved the lives of a crew of scurvy-ridden French seamen in the 16th century. The tree was introduced into Europe at this time, and was planted in Paris in 1536. Northern White-cedars are slow-growing but quite long-lived.

Identification   Height: 40–70′; diameter: 1–3′. Medium-size evergreen with much-branched trunk and conical crown of short, spreading branches. Leaves yellow-green above, paler blue-green below; tiny (1/16–1/8″ long) and scalelike; opposite, in 4 rows. Cones light brown, elliptical, 3/8″ long; growing upright from short stalks, with 8–10 leathery cone-scales.

Habitat   Swampy areas and limestone uplands with neutral or alkaline soils; often in pure stands.

Range   SE. Manitoba through Great Lakes area and much of S. Ontario and Quebec to Maine and New York; local farther south; to 3000′ in the South.

158

Norway Spruce *Picea abies*

This large conifer has the familiar "Christmas-tree" shape and big, showy cones. Introduced from northern Europe, where it is an important timber species, the Norway Spruce is widely planted in North America. Many cultivated varieties have been developed for ornament. The timber is used for pulpwood, in cabinetry, and in making the sounding boards of violins.

Identification   Height: 80′; diameter: 2′. Tall, pyramid-shaped evergreen with straight trunk and spreading branches. Needles shiny dark green, marked with whitish lines; ½–1″ long, with sharp tips; growing on all sides of twig. Cones light brown, cylindrical; 4–6″ long, and growing downward from twig; numerous cone-scales have sharp points and are sometimes toothed; cones open and shed 1 year after maturing.

Habitat   Cool, humid, temperate regions with moist soils.

Range   SE. Canada and NE. United States, often at high altitudes; native to N. and central Europe.

Red Spruce *Picea rubens*

While this is the only spruce that does not occur in the West, it reaches farther south than any other. A handsome shade tree, it is quite similar to the Black Spruce (*P. mariana*), with which it grows in the northern part of its range; the Black Spruce can then be distinguished by its smaller, dull gray cones and often irregular or spreading crown. The Red Spruce is also called Eastern Spruce or Yellow Spruce.

Identification Height: 50–80′; diameter: 1–2′. Medium-size to tall evergreen with conical crown. Needles shiny green with whitish lines; 1–2″ long, with sharp points; growing from all sides of twig. Cones reddish brown, cylindrical; 1¼–1½″ long; growing downward from short stalks; often with fine teeth on stiff cone-scales; falling at maturity.

Habitat Mountain areas with rocky soils; often in pure stands.

Range Ontario to Nova Scotia and New England; south in mountains to E. Tennessee and W. North Carolina; to 4500–6500′ in the South.

162

Balsam Fir *Abies balsamea*

This is the only fir native to the Northeast. Its aromatic needles perfume the air of northern coniferous forests; these same needles are used to make fragrant balsam pillows. In winter, the foliage becomes an important food source for a variety of large mammals, notably deer and moose.

Identification Height: 40–60′; diameter: 1–1½′. Medium-size evergreen with sharply pyramidal crown and spreading branches. Needles shiny dark green above and below; lower surface marked with 2 distinctive white bands; flat, sometimes with rounded tip; 1–1½″ long, growing in 2 nearly perpendicular rows on twigs. Cones dark purplish brown; cylindrical, 2–3¼″ long; growing upright from upper twigs; cone-scales have fine hairs.

Habitat Coniferous forests; often in pure stands.

Range W. Alberta east to Newfoundland, south to Minnesota, NE. Iowa, and Pennsylvania; to timberline in the North; above 4000′ in southern mountains.

Eastern Hemlock *Tsuga canadensis*

This is a popular, elegant member of a group of graceful evergreens. Hemlock trees are not related to the herb that was used to poison Socrates; in fact, European settlers to this country made tea from the young leaves of Eastern Hemlock; the needles were also once an ingredient in old-fashioned root beer.

Identification Height: 60–70′; diameter: 2–3′. Large evergreen with conical crown; branches slender; horizontal or often drooping; tip of tree (leader) often curved and drooping. Needles shiny dark green above and below, with 2 narrow white bands on lower surface; flat, flexible; ⅜–⅝″ long, and growing in 2 rows. Cones small, brown, elliptical; ⅝–¾″ long; growing downward from ends of twigs; with rounded cone-scales.

Habitat Cool valleys and slopes with moist, acid soils; often in pure stands.

Range E. Minnesota to Cape Breton Island; south in mountains to N. Alabama; to 3000′ in the North and 2000–5000′ in the South.

White Spruce *Picea glauca*

With its close relative the Black Spruce (*P. mariana*), this species forms vast forests across northern North America. At treeline, the White Spruce often occurs as a small shrub. In winter, the needles provide food for a variety of wildlife, including grouse, ptarmigans, deer, and rabbits.

Identification — Height: 40–100'; diameter: 1–2'. Medium-size to tall evergreen with horizontal branches and a conical crown. Needles blue-green with whitish lines; stiff, with sharp points; ½–¾" long, growing chiefly from upper side of twigs. Cones shiny light brown; cylindrical, 1¼–2½" long; with thin cone-scales; growing downward from ends of twigs; falling at maturity.

Habitat — Coniferous forests with many types of soil; in pure stands or with other conifers, often with Black Spruce.

Range — From northern limit of trees in Alaska, N. Quebec, and Newfoundland south to British Columbia, Great Lakes, and Maine; to timberline (2000–5000') in mountains.

Tamarack *Larix laricina*

Although the Tamarack is a conifer and looks like an evergreen, it sheds its needles every autumn. The wood is very sturdy and is useful for large projects, such as house frames, railroad ties, and fence posts. Also known as the Eastern Larch, this species is a popular ornamental in northern areas because it is well adapted to cold.

Identification Height: 40–80'; diameter: 1–2'. Medium-size tree with conical crown of open, horizontal branches and thin, straight trunk. Needles light blue-green, soft, slender; ¾–1" long and ¹⁄₃₂" wide; in clusters or scattered along the twig on leaders; turning yellow in fall. Cones bright red-brown, turning darker brown with rounded cone-scales; ½–¾" long; growing upright and falling in second year.

Habitat Bogs and swamps with peaty soils; in uplands with loamy soils; often in pure stands.

Range Alaska and N. British Columbia to Labrador, south to Minnesota and N. New Jersey; local farther south; near sea level in the North; to 4000' in southern part of range.

170

Scotch Pine *Pinus sylvestris*

Introduced from Europe, where it is widespread, the Scotch Pine has become a favorite ornamental throughout North America. This species is the most widespread pine in the world, and in Europe it has long been an important timber species; it is also a valuable source of turpentine. The blue-green foliage and warm, reddish bark make the Scotch Pine a very picturesque species.

Identification   Height: 70'; diameter: 2', often larger in old trees. Tall evergreen with conical crown when young, becoming rounded and irregular. Needles blue-green, stiff, becoming twisted; 1½–2¾" long, in bundles of 2. Cones yellowish-brown, egg-shaped; 1¼–2½" long, with thin, flat cone-scales; opening at maturity.

Habitat   Sandy to loamy soils; forested areas, suburbs, gardens, and cities.

Range   Naturalized locally from SE. Canada to Iowa, east to New England; native to Europe and Asia.

172

Red Pine *Pinus resinosa*

This common tree is used throughout its range as an ornamental. It is also known as the Norway Pine, perhaps because it has been confused with the Norway Spruce (*Picea abies*), an introduced conifer planted in some of the same areas. Each year the Red Pine adds another row of branches to its lengthening trunk.

Identification Height: 70–80′; diameter: 1–3′. Large evergreen with irregular, broad crown (sometimes rounded) and spreading branches. Needles dark green, slender, 4¼–6½″ long; in bundles of 2. Cones shiny light brown, egg-shaped, 1½–2¼″ long, with keeled cone-scales that lack prickle at tip; opening and falling soon after maturity.

Habitat Sandy plains and other areas with well-drained soils; usually in mixed forests.

Range SE. Manitoba and NE. Minnesota to Nova Scotia and Pennsylvania; local in E. West Virginia and some other areas; 700–1400′ in northern part of range; to 2700′ in Adirondack Mountains; 3800–4300′ in West Virginia.

Eastern White Pine *Pinus strobus*

This is the largest conifer in eastern North America; it once dominated the forests of the Northeast, and in a few remote, protected places pure stands of huge old trees can still be found. The durable wood has a variety of uses, especially in construction and for pulpwood. This species was introduced to England in the early 17th century; in that country it is known as the Weymouth Pine. The White Pine is the state tree of Maine.

Identification Height: 100' or taller; diameter: 3–4'. Tall evergreen with conical crown in young trees, becoming irregular; branches horizontal. Needles blue-green, slender; 3–4" long, in bundles of 5. Cones yellow-brown, narrowly cylindrical; 4–8" long with thin, rounded cone-scales.

Habitat Well-drained areas with sandy soils; sometimes in pure stands.

Range SE. Manitoba and Minnesota east to Newfoundland and New England, south to NE. Iowa and N. Georgia; sea level to 2000'; to 5000' in southern Appalachians.

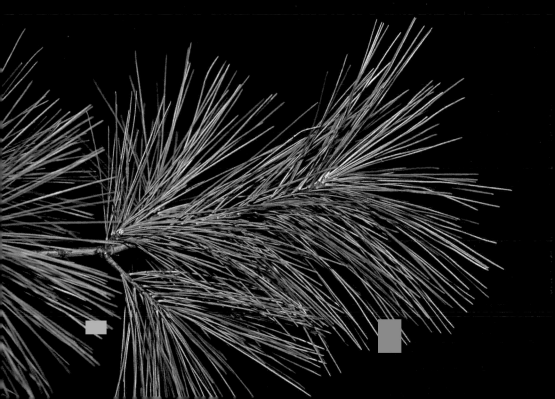

Longleaf Pine *Pinus palustris*

This abundant pine is one of the most important timber trees of the Southeast. Its wood is used for construction lumber, pulpwood, and pilings. Longleaf Pine is also a leading source of turpentine and resin. This tree has the longest needles and the largest cones of any pine in the East; young trees, especially, have very long needles, often reaching 18″.

Identification   Height: 80–100′; diameter: 2–2½′. Tall tree with irregular open crown and spreading branches. Needles dark green, long, and drooping; somewhat stout but flexible; 10–15″ long, in bundles of 3 and crowded along twigs. Cones brown; cylindrical or conical, 6–10″ long; cone-scales have keel and small prickle or bristle at tip; opening and shedding at maturity.

Habitat   Sandy, well-drained soils on flatlands; coastal areas; often in pure stands.

Range   SE. Virginia south along coastal plain to central Florida, west to E. Texas; to 600′ in most areas; to 2000′ in southern foothills of Appalachians.

178

Guide to Families

"What tree is that?" To the beginner, learning to distinguish among the hundreds of tree species in North America may seem a formidable task. All species of trees belong to a genus and a family, however, and learning to recognize the broad, shared features of these groups is a good shortcut to identification. By learning the traits of major families, you can be sure of pleasure when you go for a walk in the woods or a stroll through a city park.

Pines
All members of the pine family have needlelike leaves and woody cones that bear the seeds; all but the larches are evergreen. Within the family, different genera have other identifying characteristics: Pines (*Pinus*) have needles in clusters of two to five; spruces (*Picea*) have sharp, pointed individual needles; firs (*Abies*) have upright cones clustered near the top of the crown, and larches (*Larix*) have deciduous needles in pincushionlike collections on short spur shoots.

Cypresses
The cypress (or "cedar") family provides the other major group of evergreens or conifers in the East. Two genera, arborvitae (*Thuja*) and white-cedar (*Chamaecyparis*), have scalelike leaves attached so closely that the twig is completely covered, and small woody cones. Junipers (*Juniperus*), on the other hand, have awl-like leaves and

180

berrylike cones; despite its misleading common name, the abundant Eastern Redcedar—the most common juniper in eastern North America—can be recognized as a juniper by its "berries."

Redwoods
The redwoods are represented in the East by a single, distinctive genus, baldcypress (*Taxodium*). Another of the deciduous "evergreens," baldcypress is easily recognized by the swampy habitats it occupies, the flared and fluted lower trunks, and the "knees" that protrude above the surrounding water.

Beeches
There are many and distinctive families of broadleaf or hardwood trees. One of the most important is the beech family, which includes beeches (*Fagus*) and the true chestnut (*Castanea*) as well as the oaks (*Quercus*). All members of the family have a simple (undivided) leaf and a nut enclosed in some kind of husk. Deciduous oaks have lobed leaves; the leathery evergreen leaves of live oaks are not lobed, and some are willowlike.

Willows
The willow or poplar family is an important group of fast-growing trees often found in moist or wet habitats. All are deciduous and bear their flowers in catkins (pussy willows are an example) but the leaves of willows (*Salix*)

tend to be long and narrow, while those of poplars and aspens (*Populus*) are typically broad and triangular.

Birches and Walnuts The birch family includes birches (*Betula*), alders (*Alnus*), hornbeams (*Carpinus*), and hophornbeams (*Ostrya*); its members have simple deciduous leaves and bear their flowers in catkins. Many birches have distinctive, pale, peeling bark. The walnut family includes walnuts (*Juglans*) and hickories (*Carya*). The group is distinguished by large, pinnately compound leaves and a large nut enclosed within a tough, fleshy husk. Walnuts have many leaflets (up to 23) on a leaf, and the husk is not divided into segments. Hickories typically have five to nine leaflets and husks with distinctive segments; the Pecan is a hickory.

Maples and Elms A large and generally easily recognized family, the maples (*Acer*) have a familiar palmately lobed leaf that is seen on the Canadian flag; the distinctive winged seeds are joined at the base in pairs, forming a double key, or samara. The pinnately compound leaf of the Boxelder differs from the classic "maple leaf," but the species still has opposite branching and bears the winged maple fruit. The elm family includes elms (*Ulmus*) and hackberries (*Celtis*). Both have simple, alternate leaves, highly asymmetrical at the base. Elms have small seeds completely surrounded by a wing—another kind of key.

Magnolias   The magnolia family is known for its large, showy flowers; many species also have very large leaves. The Yellow-poplar (*Liriodendron*) is distinctive for its tall, straight growth form. It also belongs to the magnolia family and has a conelike aggregation of fruits.

Witch-hazels and   With its star-shaped leaf and golf-ball size, pendent,
Sycamores   woody fruit, the Sweetgum is the most distinctive member of the witch-hazel family. The shrubby Witch-hazel is itself most easily recognized by the fall-borne yellow flowers with their straplike petals. The sycamore family includes only a single genus, *Platanus*. Older Sycamores are easily recognized by the smooth, light (often white) inner bark that is revealed on the upper branches and boles. Like the Sweetgum, the Sycamore has a golf-ball size fruit that hangs from a slender stem, but the palmate leaves are not star-shaped.

Roses   This large family includes many genera important as crop plants and ornamentals—cherry (*Prunus*), apple (*Malus*), pear (*Pyrus*), mountain-ash (*Sorbus*), serviceberry (*Amelanchier*), and hawthorn (*Crataegus*).

This diverse family shows great variety in leaf, flower, and fruit characteristics. Attractive flowers and fleshy, edible fruits are common, however. The apples and their allies bear a pome (a large, fleshy fruit with seeds in a papery chamber); the plums and cherries have a fleshy, one-seeded fruit (or drupe) that does not split open.

Legumes and Cashews The pea or legume family includes a very large number of flowers and crops, as well as some important trees. In eastern North America, the Honeylocust, Blacklocust, and Kentucky Coffeetree are easily recognized by their pinnately compound leaves, thorns, and the fruit, which is a legume, a brown or black papery version of a pea pod with hard, dark seeds. The cashew family is mostly tropical but includes the sumacs (*Rhus*), which are distinctive for their pinnately compound leaves and are typically bright scarlet in the fall. A related genus, *Toxicodendron*, includes toxic species such as poison ivy.

Hollies This family includes 300 or more species worldwide, but most are tropical; only 16 are common in North America. Hollies are small to medium-size; red berries and small, spiny evergreen leaves characterize many species, but others lack the spine-tipped leaves and have dark fruit.

Buckeyes and Dogwoods Members of the buckeye, or horsechestnut, family are easily recognized by their large, palmately compound leaves, conspicuous upright shoots of white flowers, and large nuts enclosed in a leathery capsule. Dogwoods have opposite branches and a one-seeded fleshy fruit. This variable family is best known for those species in the genus *Cornus* with showy flowers (in which the "petals" are actually modified leaves). The tupelos (*Nyssa*), also in the dogwood family, are characteristic of moist to swampy habitats; they often have swollen trunks.

Heaths and Olives The heath family contains a rich variety of shrubs and trees. Leathery evergreen leaves are characteristic of many genera in this family, as are very showy flowers, like those of mountain-laurel (*Kalmia*) and rhododendron (*Rhododendron*). The olive family is best represented by the ash trees (*Fraxinus*), with their opposite, pinnately compound, deciduous leaves and seed with an elongated terminal wing (another type of samara). Ashes grow in moist habitats, such as bottomlands and streamsides.

As you see, many trees can be generally identified with just a few family characteristics, thereby providing an answer to the question, "What tree is that?" Don't be discouraged by exceptions to any rule of identification!

Glossary

Alternate
Single along a twig or shoot; not whorled or in pairs.

Catkin
A compact cluster of reduced, stalkless, and usually unisexual flowers.

Cone-scale
One of the overlapping, seed-bearing scales of a cone.

Crown
The branches, twigs, and leaves at the top of a tree.

Deciduous
Shedding leaves seasonally, leafless for part of the year.

Drupe
A fleshy fruit with a central, stonelike core containing 1 or more seeds.

Genus
A group of closely related species. Plural, genera.

Introduced
Established in an area by man; exotic or foreign.

Opposite
In pairs along a twig or shoot, with 1 on each side; not alternate or whorled.

Persistent
Remaining attached, and not falling off.

Sheath
In some conifers, the papery tube enclosing the bases of needles.

Shrub
A woody plant, smaller than a tree and with several stems arising from a single base.

Species
A group of plants or animals composed of individuals that interbreed and produce similar offspring.

Tree line
The upper limit of tree growth at high latitudes or on mountains; timberline.

Wing
A thin, flat, dry projection on a fruit or seed.

Index

The Audubon Society

The National Audubon Society is among the oldest and largest private conservation organizations in the world. With over 560,000 members and more than 500 local chapters across the country, the Society works in behalf of our natural heritage through environmental education and conservation action. It protects wildlife in more than eighty sanctuaries from coast to coast. It also operates outdoor education centers and ecology workshops and publishes the prizewinning AUDUBON magazine, AMERICAN BIRDS magazine, newsletters, films, and other educational materials. For further information regarding membership in the Society, write to the National Audubon Society, 950 Third Avenue, New York, New York 10022.